Mercury

The Enigma of Extremes

JD ARDEN

Preface: First Among Planets

To call Mercury the "first planet" is to speak both literally and symbolically. Closest to the Sun, it occupies the most extreme and precarious position in the solar system—a world scorched by the Sun's rays and frozen in its shadow. It is a planet of contradictions, where searing heat and freezing cold coexist, and where a seemingly barren landscape holds secrets that defy expectations.

For all its proximity to Earth, Mercury remains one of the least understood planets. It has no breathable atmosphere, no comforting seas, no thick clouds to hide its surface. What it offers instead is stark beauty, a pockmarked surface shaped by billions of years of impacts and the relentless pull of the Sun's gravity. It is a reminder of what lies at the extremes of existence—a world unprotected, exposed, and enduring.

Mercury's small size and close orbit have often relegated it to the background of planetary science, overshadowed by Mars's red deserts and Venus's mysterious clouds. Yet, Mercury is no less remarkable. Its density hints at a history of cosmic violence, its magnetic field defies conventional wisdom, and its polar craters, bathed in perpetual shadow, harbor ice—an improbable discovery on such a scorched world.

This book seeks to bring Mercury out of the Sun's shadow, to explore the paradoxes of a planet that both reveals and resists understanding. Through science, mythology, and philosophy, we will journey to the edge of the Sun's domain, where Mercury orbits like a silent sentinel, bearing witness to the extremes of our solar system.

Chapter 1: Mercury's Harsh Realities

Mercury, the smallest and closest planet to the Sun, is a world of relentless extremes. It is a place where the ordinary rules of planetary behavior are bent and where survival, in any conventional sense, is inconceivable. To understand Mercury, one must strip away preconceptions about what a planet should be and confront the raw, unfiltered reality of a world shaped by extremes of heat, cold, and proximity to the Sun.

Mercury's defining feature is its extreme environment, a consequence of its closeness to the Sun. At midday, surface temperatures soar to a blistering 430 degrees Celsius, hot enough to melt lead and vaporize most materials. But this intense heat is not constant; it is replaced by bone-chilling cold when the Sun sets. Without an atmosphere to trap heat, nighttime temperatures plunge to a frigid minus 180 degrees Celsius, creating a daily temperature swing of over 600 degrees—the largest in the solar system.

This dramatic contrast is a direct result of Mercury's lack of a significant atmosphere. While Earth's thick air moderates temperatures and protects the surface from the vacuum of space, Mercury has only a thin exosphere, composed of atoms blasted off its surface by solar winds. This exosphere, far too insubstantial to provide insulation, leaves the planet exposed to the full intensity of solar radiation and the bitter cold of space.

Mercury's slow rotation adds another layer to its harshness. A single day—defined as one complete rotation on its axis—lasts 59 Earth days. Coupled with its fast orbit around the Sun, which takes just 88 Earth days, this creates a strange rhythm. A single day-night cycle on Mercury lasts 176 Earth days. The Sun lingers in the sky for weeks before setting, only to be followed by weeks of darkness. This prolonged exposure to both sunlight and shadow amplifies the planet's extremes, making it a world where time itself feels stretched.

The surface of Mercury is equally unforgiving, scarred by billions of years of impacts. Unlike Earth, where weather and tectonic activity erase craters over time, Mercury's lack of atmosphere and internal activity has preserved its surface almost in its entirety. It is a fossil record of the

Mercury

solar system's violent past, marked by craters ranging from tiny pockmarks to vast basins.

The most striking of these is the Caloris Basin, a massive impact crater over 1,550 kilometers wide. Formed by a collision with an asteroid or comet during the early days of the solar system, the basin is surrounded by chaotic terrain created by the shockwaves of the impact. Standing on Mercury's surface, one could trace these scars and read the story of a planet shaped by relentless bombardment.

Yet Mercury is not as simple as it seems. Beneath its heavily cratered surface lies one of the most enigmatic features of the planet: its unusually large iron core. Accounting for about 85% of the planet's radius, this core is proportionally larger than that of any other planet in the solar system. It makes Mercury the second densest planet after Earth, raising profound questions about its origin and evolution.

Why does Mercury have such a massive core? One theory suggests that it was once much larger but lost much of its outer mantle in a catastrophic collision early in its history. Another possibility is that the intense heat and solar winds of the young Sun stripped away Mercury's lighter elements during its formation, leaving behind a dense metallic core. Either explanation points to a history of violence and survival, a story of a planet shaped by its proximity to the Sun and its resilience in the face of cosmic forces.

The planet's surface also hints at ongoing geological processes. Mercury is not entirely static; its surface shows evidence of contraction. As its core cools and solidifies, the planet has shrunk, creating fault lines and scarps that crisscross the terrain. Some of these scarps rise hundreds of meters and stretch for hundreds of kilometers, resembling the wrinkles on an aging face. This slow contraction is a reminder that even the smallest, most ancient planets remain dynamic, their stories still unfolding.

Despite its proximity to Earth, Mercury has been one of the most challenging planets to study. The Sun's immense gravitational pull makes it difficult for spacecraft to reach Mercury and enter orbit without expending enormous amounts of energy. For centuries, much of what we knew about Mercury came from Earth-based telescopes, which could only

Mercury

provide fleeting glimpses of the planet during its closest approaches to Earth.

The breakthrough came in the mid-20th century with the advent of space exploration. NASA's Mariner 10 spacecraft, launched in 1973, became the first to visit Mercury, capturing images of its surface and revealing its cratered, moon-like appearance. Decades later, the MESSENGER mission (2004–2015) transformed our understanding of Mercury, mapping its entire surface and uncovering new mysteries, such as the presence of water ice in its shadowed craters.

Even with these advancements, much about Mercury remains unknown. Its density, its magnetic field, and its geological history continue to challenge scientists, forcing them to rethink traditional models of planetary formation and evolution.

Mercury's harsh realities are a testament to the extremes of existence. It is a planet stripped to its essence, unadorned by the complexities of atmosphere or lush ecosystems. Yet, in its starkness lies a profound beauty. Mercury endures, orbiting closer to the Sun than any other planet, its surface battered but unbroken, its core a relic of a bygone era.

To study Mercury is to confront the raw power of the solar system and the resilience of matter in the face of adversity. It is to understand that even in the harshest conditions, there is a story of survival, a story of endurance. Mercury's realities are not just scientific curiosities; they are lessons in the nature of extremes, the persistence of worlds, and the boundless creativity of the cosmos.

Chapter 2: Ice in the Inferno

Mercury's harshness is almost legendary. Its searing daytime temperatures and freezing nights have earned it a reputation as one of the most extreme planets in the solar system. It seems a place utterly inhospitable, incapable of sustaining anything but the sheer brute force of physics. Yet, in 2012, a discovery shattered this perception: Mercury, the planet closest to the Sun, harbors ice.

The revelation of water ice on Mercury challenged everything we thought we knew about this scorched world. Found in the permanently shadowed craters at the planet's poles, this ice survives in some of the coldest places in the solar system. These craters, shielded from sunlight by their steep walls, remain in perpetual darkness, their temperatures plunging to minus 200 degrees Celsius—cold enough to preserve ice for billions of years.

The ice was first suspected during the 1990s when Earth-based radar observations revealed bright reflections from Mercury's poles. At the time, these reflections were intriguing but not conclusive. It wasn't until NASA's MESSENGER spacecraft entered orbit around Mercury in 2011 that scientists could confirm the presence of water ice. Using a combination of neutron spectrometry, laser altimetry, and high-resolution imaging, MESSENGER detected the unmistakable signature of frozen water in the polar regions.

How, then, does a planet so close to the Sun retain ice? The answer lies in the unique geometry of Mercury's poles. Because Mercury has almost no axial tilt—its rotational axis is nearly perpendicular to its orbital plane—the Sun never rises high enough to illuminate the floors of certain craters near the poles. These shadowed regions, untouched by sunlight for billions of years, become cold traps, preserving volatile materials like water ice that are otherwise vaporized by the Sun's heat.

The ice likely originates from two main sources: cometary impacts and the solar wind. Comets, which are rich in water and other volatile compounds, have bombarded Mercury throughout its history, delivering material that could have accumulated in these cold traps. Additionally, the solar wind—streams of charged particles emitted by the Sun—carries hydrogen ions that can combine with oxygen on Mercury's surface to form water molecules. These molecules, once freed by impacts or surface interactions, may migrate to the poles, where they become trapped in the shadows.

But the ice is not just frozen water; it is mixed with organic materials, dark, carbon-rich compounds that give it a distinct appearance. These organics, likely delivered by the same comets that brought the water, are intriguing for their potential connection to the building blocks of life. Although Mercury is far too hostile to sustain life, the presence of these materials raises questions about the distribution of organic compounds in the early solar system and their role in seeding planets like Earth.

The discovery of ice on Mercury also raises profound questions about planetary evolution and survival in extreme conditions. Why has this ice persisted for billions of years? What does its presence tell us about the processes that shape planets at the edges of habitability?

One clue lies in the nature of Mercury's environment. Despite its proximity to the Sun, Mercury's lack of atmosphere means there is no weather to erode or redistribute the ice. The planet's geological inactivity further ensures that the ice remains undisturbed, locked away in its shadowed refuges. In many ways, these craters serve as time capsules, preserving not only water but also a record of the solar system's volatile history.

The implications of this discovery extend beyond Mercury. It suggests that even in the most hostile environments, pockets of stability can exist. These cold traps, isolated from the planet's extremes, offer a glimpse into the resilience of matter and the persistence of water in the harshest conditions. They remind us that the universe often defies expectations, creating unlikely havens in the midst of chaos.

For scientists, Mercury's polar ice is a treasure trove of data waiting to be explored. Future missions, such as the joint European Space Agency and JAXA BepiColombo spacecraft, aim to investigate these regions in greater detail, unraveling the composition, origin, and history of the ice. Such studies could provide new insights into the distribution of water across the solar system, from the Moon's shadowed craters to the icy moons of Jupiter and Saturn.

Mercury

Philosophically, the discovery of ice on Mercury challenges our perceptions of extremes. We often think of harsh environments as devoid of potential, places where life and its prerequisites cannot exist. Yet, Mercury's ice reminds us that extremes can coexist with pockets of preservation, that even in the blazing heat of a planet bathed in sunlight, there are shadows where cold and stillness reign.

Mercury's ice also serves as a metaphor for resilience. It is a substance that has endured unimaginable odds, surviving on a world where nothing else can. Its presence is a testament to the complexity of planetary systems and the surprising ways in which nature preserves its secrets.

To look at Mercury and find ice is to be reminded that the universe is full of surprises. It is a reminder that the boundaries of possibility are not fixed but ever-expanding, shaped by discovery and imagination. The ice in Mercury's craters, gleaming in the perpetual shadow, is more than frozen water. It is a symbol of endurance, a relic of the solar system's past, and a promise of mysteries yet to be uncovered.

Chapter 3: The Messenger's Journey

Mercury's mysteries have always been difficult to uncover. Its proximity to the Sun, small size, and inhospitable conditions make it one of the most challenging planets to study. For much of human history, Mercury remained a distant enigma, a faint point of light visible only during twilight, its surface forever hidden from Earthbound telescopes. But in the latter half of the 20th century, the age of space exploration ushered in a new era, and Mercury finally began to reveal its secrets.

The first major breakthrough came with **Mariner 10**, a pioneering spacecraft launched by NASA in 1973. Using gravitational slingshots—at the time, a novel concept—Mariner 10 flew past Venus and used its gravity to alter its trajectory toward Mercury. In 1974, the spacecraft became the first human-made object to visit Mercury, providing humanity with its first close-up images of the planet's cratered surface.

What Mariner 10 revealed was unexpected: Mercury was not smooth or featureless as many had imagined but rugged and scarred, its surface resembling a smaller, denser version of Earth's Moon. The spacecraft photographed about 45% of the planet, uncovering vast craters, cliffs, and plains. Among the most notable discoveries was the **Caloris Basin**, an enormous impact crater surrounded by concentric rings of mountainous terrain. The basin's sheer scale—over 1,500 kilometers wide—was a stark reminder of the violent collisions that shaped Mercury's early history.

While Mariner 10's mission provided invaluable data, it also left many questions unanswered. The spacecraft only conducted three flybys of Mercury, and its observations were limited to the same hemisphere of the planet each time. For decades, scientists were left to speculate about the unseen half of Mercury, its geology, and the forces that shaped it.

In 2004, NASA launched the **MESSENGER mission** (MErcury Surface, Space ENvironment, GEochemistry, and Ranging), a spacecraft designed to answer these lingering questions. MESSENGER faced immense challenges. To reach Mercury, it had to execute six gravity assists—one with Earth, two with Venus, and three with Mercury itself—before finally settling into orbit in 2011. This intricate path, covering over 7.9 billion

Mercury

kilometers, took seven years to complete, a testament to the complexity of traveling to a planet so close to the Sun.

MESSENGER's orbital mission was nothing short of transformative. Over its four years circling Mercury, the spacecraft mapped the planet's entire surface, revealing a world far more dynamic and complex than previously imagined. Among its many discoveries, MESSENGER confirmed the presence of water ice in Mercury's polar craters, a finding that upended our understanding of the planet's ability to retain volatile materials.

MESSENGER also detected traces of organic compounds in the polar regions, hinting at the delivery of these materials by comets and asteroids. This discovery connected Mercury's history to that of the broader solar system, showing how even a scorched world near the Sun could preserve evidence of its violent and interconnected past.

Another surprising finding was the planet's magnetic field. Mariner 10 had detected this field decades earlier, but MESSENGER revealed its true nature: Mercury's magnetic field is weak, offset from its center, and asymmetrical, with a stronger presence in the northern hemisphere than in the southern. This magnetic irregularity challenges traditional models of planetary dynamos, the processes by which planetary cores generate magnetic fields.

MESSENGER's instruments also measured Mercury's surface composition, revealing an abundance of sulfur and other volatile elements. This composition suggests that Mercury formed under conditions very different from those of Earth or Venus, adding new layers of complexity to our understanding of planetary formation.

The mission also uncovered evidence of Mercury's geological activity. While the planet lacks plate tectonics, its surface is marked by scarps and cliffs—evidence that it has contracted over time as its core cooled and solidified. Some of these features are geologically recent, suggesting that Mercury's surface is still evolving, albeit slowly.

The data from MESSENGER not only deepened our understanding of Mercury but also raised new questions. Why does a planet so small retain a magnetic field? How has its surface remained geologically active for so

Mercury

long? And what do its composition and history reveal about the early solar system?

As MESSENGER's fuel dwindled, its mission came to a dramatic end. In 2015, the spacecraft was intentionally crashed into Mercury's surface, creating a new crater and marking the end of an era in Mercury exploration. Yet, its legacy lives on in the vast troves of data it collected, data that scientists continue to analyze in search of answers.

The next chapter in Mercury exploration is already unfolding. The **BepiColombo mission**, a joint effort between the European Space Agency (ESA) and the Japan Aerospace Exploration Agency (JAXA), launched in 2018 and is currently en route to Mercury. Expected to arrive in 2025, BepiColombo will carry two orbiters equipped with advanced instruments designed to study the planet's magnetic field, surface composition, and exosphere in unprecedented detail.

BepiColombo represents the continuation of a journey that began with Mariner 10 and MESSENGER, a journey to uncover the secrets of a planet that has long been overshadowed by its larger, more charismatic neighbors. Each mission builds on the discoveries of the last, adding new pieces to the puzzle of Mercury's enigmatic nature.

The story of Mercury's exploration is one of perseverance and ingenuity. It is a testament to humanity's ability to overcome immense challenges in the pursuit of knowledge. Mercury, with its harsh realities and extreme environment, stands as a symbol of the universe's capacity to defy expectations. Through missions like MESSENGER and BepiColombo, we are reminded that even the most inaccessible worlds have stories to tell—stories that deepen our understanding of the cosmos and our place within it.

Chapter 4: Mercury's Magnetic Surprise

Mercury is a planet of contradictions, and its magnetic field is among the most unexpected and fascinating of these. For decades, scientists assumed that such a small, barren, and ancient world, battered by the Sun's intense heat and radiation, would lack any significant magnetic activity. After all, planets like Mars and Venus—larger and seemingly more dynamic in their histories—have no global magnetic fields. And yet, Mercury defies this expectation, possessing a magnetic field that challenges long-held assumptions about planetary interiors and evolution.

The discovery of Mercury's magnetic field was one of the most surprising results of the **Mariner 10** mission in the 1970s. As the spacecraft flew past Mercury, its magnetometer detected a weak but unmistakable global magnetic field, one that resembled Earth's in its dipole structure, with north and south magnetic poles. This finding was puzzling. How could a planet so small—less than half Earth's diameter—sustain a magnetic field when larger planets like Mars could not?

The magnetic field was revisited decades later with the **MESSENGER mission**, which orbited Mercury from 2011 to 2015. MESSENGER's instruments provided a far more detailed picture, revealing that Mercury's magnetic field is both weak and asymmetric. It is about 100 times weaker than Earth's and offset significantly from the planet's center, with the northern hemisphere experiencing stronger magnetic forces than the southern. This lopsided nature added a new layer of complexity to an already enigmatic phenomenon.

At the heart of this mystery is the question of how planetary magnetic fields are generated. Magnetic fields arise from a process called the **dynamo effect**, which occurs in a planet's liquid core. As molten metal in the core moves and rotates, it generates electric currents, which in turn produce magnetic fields. For this dynamo process to occur, a planet needs both a sufficiently large, molten core and enough energy to keep it in motion.

Mercury

Mercury's size makes this unlikely. Small planets cool quickly, and conventional wisdom suggested that Mercury's core should have solidified long ago, halting any dynamo activity. Yet, MESSENGER's findings confirm that at least part of Mercury's core remains liquid, allowing the dynamo effect to persist.

One explanation for this lies in Mercury's composition and proximity to the Sun. Mercury has an unusually large iron core, making up about 85% of the planet's radius. This high iron content gives the core a significant heat reservoir, allowing it to stay partially molten despite the planet's small size. Furthermore, the intense tidal forces exerted by the Sun may contribute to the core's motion, helping to sustain the dynamo.

But the story does not end there. Mercury's magnetic field is not just a scientific curiosity; it also plays a crucial role in shaping the planet's interaction with the Sun. Mercury is bathed in the solar wind—a constant stream of charged particles emitted by the Sun. Without a magnetic field, these particles would bombard the planet's surface unimpeded, stripping away its tenuous exosphere and eroding its surface.

Mercury's magnetic field, while weak, provides a degree of protection by deflecting some of these particles. However, its offset nature means that the planet's southern hemisphere is less shielded than the north, leading to differences in surface erosion and particle deposition. This asymmetry has created a dynamic and ever-changing environment, where Mercury's surface and exosphere are constantly being reshaped by the Sun's influence.

One of the more intriguing aspects of Mercury's magnetic field is its ancient origin. MESSENGER detected evidence of magnetized rocks on Mercury's surface, suggesting that the planet's magnetic field has existed for billions of years. These ancient rocks indicate that Mercury's dynamo was once much stronger than it is today, raising questions about how and why the magnetic field weakened over time.

The persistence of Mercury's magnetic field also challenges our understanding of planetary evolution. Why does Mercury, a small and seemingly inert world, maintain a magnetic field while Mars, a larger planet, does not? What does Mercury's magnetic activity reveal about the conditions required for a dynamo to operate?

Mercury

Scientists are only beginning to unravel these questions. The ongoing analysis of MESSENGER's data, combined with future observations from the **BepiColombo mission**, promises to shed new light on the mechanisms behind Mercury's magnetic field. These studies could have implications far beyond Mercury, offering insights into the magnetic histories of other planets and even exoplanets in distant star systems.

Mercury's magnetic field is not just a scientific anomaly; it is a testament to the planet's resilience. Despite its small size, extreme environment, and apparent simplicity, Mercury continues to surprise us with its ability to sustain processes that defy expectations. Its magnetic field, weak and lopsided though it may be, is a reminder that even the smallest worlds can hold immense secrets.

Philosophically, Mercury's magnetic field challenges our assumptions about limits. We often equate smallness with fragility, assuming that only the largest and most robust systems can endure. Yet, Mercury shows that persistence and complexity are not dictated by size. Its magnetic field, though diminished, remains active—a quiet defiance of the forces that seek to extinguish it.

To study Mercury's magnetism is to confront the dynamism of a planet that refuses to conform to expectations. It is a journey into the depths of planetary interiors, a search for the unseen forces that shape worlds, and a reminder that the universe is full of surprises, even in the places we least expect.

Chapter 5: Mercury and the Early Solar System

To understand Mercury is to journey back in time to the dawn of the solar system, over 4.6 billion years ago, when a swirling cloud of gas and dust began to collapse under the force of gravity. From this cosmic maelstrom, the Sun was born, and around it formed a disk of material that would eventually coalesce into the planets. Mercury, closest to the newborn Sun, emerged as a unique and enigmatic world—a planet whose origins and evolution continue to challenge our understanding of how solar systems take shape.

Mercury's proximity to the Sun makes it a key to unraveling the early history of the solar system. It is a planet stripped to its essence, a world that bears the scars of its violent formation and the relentless forces of its environment. Its composition, density, and surface features offer a glimpse into the processes that shaped the inner solar system, revealing clues about both Mercury's past and the broader dynamics of planetary formation.

One of Mercury's most striking characteristics is its high density. Despite its small size—less than 40% the diameter of Earth—Mercury is the second densest planet in the solar system. This density suggests that Mercury has an unusually large iron core, which accounts for about 85% of its radius. By comparison, Earth's core accounts for about 55% of its radius. This raises a profound question: why is Mercury so dominated by metal?

There are two leading theories to explain Mercury's composition. The first posits that Mercury initially formed as a larger planet but lost much of its outer mantle in a catastrophic collision with another planetary body. This impact would have stripped away the lighter, silicate-rich materials, leaving behind the dense metallic core and a thin crust that we see today. Such an event would make Mercury a relic of the solar system's chaotic early years, a survivor of the violent processes that shaped the planets.

Mercury

The second theory suggests that Mercury's formation was influenced by its proximity to the Sun. In the intense heat of the inner solar system, volatile elements and lighter materials may have been vaporized or blown away by the young Sun's powerful solar wind, leaving behind a planet enriched in heavy, refractory elements like iron. This explanation frames Mercury as a product of environmental selection, its composition shaped by the unique conditions of its orbit.

Both theories highlight Mercury's role as a witness to the formative chaos of the solar system. Its density, its composition, and its very existence tell a story of survival in the face of destruction, of adaptation to the harshest conditions imaginable.

Mercury's surface features further reflect its tumultuous history. The planet is heavily cratered, a testament to the relentless bombardment it endured during the **Late Heavy Bombardment**, a period about 4 billion years ago when the solar system was awash with asteroids and comets. The scars of this era are preserved on Mercury's surface, providing a fossil record of the solar system's early violence.

One of the most remarkable remnants of this era is the **Caloris Basin**, a massive impact crater over 1,500 kilometers in diameter. Formed by a collision with a large asteroid, the basin is surrounded by concentric rings of mountains and chaotic terrain created by the impact's shockwaves. On the opposite side of the planet, the force of the impact created a region of fractured and jumbled terrain, known as the "weird terrain," further evidence of Mercury's violent past.

Yet Mercury's story is not solely one of destruction. The planet's surface also reveals signs of volcanic activity, particularly in the form of vast plains that were flooded by lava billions of years ago. These volcanic flows resurfaced large areas of the planet, burying older craters and creating smoother regions. The presence of these lava plains suggests that Mercury was once more geologically active, with internal heat driving volcanic eruptions and reshaping the surface.

Even more intriguing are the planet's scarps—long, curved cliffs that crisscross the surface. These scarps are the result of Mercury's slow contraction as its core cools and solidifies. Unlike the tectonic activity seen on Earth, Mercury's scarps are a sign of a planet shrinking in on

Mercury

itself, its outer layers buckling as the interior contracts. This process, which continues to this day, makes Mercury a dynamic world despite its apparent inactivity.

Mercury's role in the early solar system extends beyond its own history. Its composition and evolution offer clues about the broader processes that shaped the inner planets. By studying Mercury, scientists gain insights into the distribution of materials in the protoplanetary disk, the effects of solar radiation on planetary formation, and the conditions that led to the diversity of planets in our solar system.

In recent years, Mercury has also become a point of comparison for exoplanets—planets orbiting stars beyond our solar system. Many of these exoplanets, particularly those in close orbits around their stars, share characteristics with Mercury, such as high densities and extreme environments. By understanding Mercury, we can begin to unlock the mysteries of these distant worlds, placing our solar system in the context of a vast and varied universe.

Mercury is more than a planet; it is a time capsule, a record of the solar system's earliest days preserved in its dense core, cratered surface, and enigmatic features. Its history is a story of survival and transformation, a story that speaks to the resilience of matter and the creativity of the cosmos.

In Mercury, we see the echoes of a time when planets were born in chaos, when collisions and catastrophes shaped the worlds we know today. It reminds us that the solar system is not a place of static perfection but a dynamic, evolving system where even the smallest worlds have roles to play.

To study Mercury is to trace the origins of the solar system, to follow the threads of its formation and see how they weave into the tapestry of the universe. It is a journey to the beginning, a search for understanding in the quiet persistence of a planet that has endured the extremes of existence.

Chapter 6: Mercury in Mythology

Before humanity understood Mercury as a small, rocky planet orbiting close to the Sun, it was a celestial wanderer, a point of light that danced across the sky in the twilight hours. Its rapid movements, visible at dawn and dusk but rarely in the depths of night, made it an object of fascination and mystery. Across cultures, Mercury's swift passage inspired myths and stories, embedding it in the collective imagination as a symbol of speed, communication, and transformation.

The name Mercury, derived from the Roman god of commerce and travel, reflects this symbolism. Mercury, the god, was a messenger—a figure who moved effortlessly between realms, delivering messages between gods and mortals, light-footed and elusive. He was associated with cunning and adaptability, traits that mirror the fleeting nature of the planet that bears his name.

The Roman Mercury was adapted from the Greek god Hermes, who shared many of the same attributes. Hermes was not only a messenger but also a guide for souls journeying to the underworld, a protector of travelers, and a patron of thieves. This duality—both a benevolent guide and a figure of mischief—reflects Mercury's own nature as a planet of extremes.

In both Greek and Roman mythology, Mercury was also linked to intellect and communication. Hermes was the god of language and writing, credited with inventing the alphabet, while Mercury was seen as a mediator, a bridge between opposites. This association resonates with the planet's position in the solar system, poised between the fiery Sun and the other, cooler worlds beyond.

In other ancient cultures, Mercury was given different names and meanings, but the themes of speed and transformation persisted. The Babylonians called it **Nabu**, after the god of writing and wisdom, who was also a messenger for the gods. The association with intellect and language aligns closely with Mercury's symbolic role in later traditions.

The ancient Egyptians associated Mercury with **Thoth**, the ibis-headed god of wisdom, writing, and magic. Thoth was a figure of balance and order, a recorder of truth and the arbiter of disputes. In this role, Mercury becomes a symbol not just of motion but of equilibrium, embodying the delicate balance required to navigate extremes—a theme that echoes Mercury's precarious existence so close to the Sun.

The planet's dual appearances in the sky, visible both at sunrise and sunset, gave rise to dual identities in some traditions. The Greeks initially named Mercury after two separate gods: **Apollo** when it appeared as the morning star and **Hermes** when it appeared in the evening. This duality reflects Mercury's role as a figure that bridges opposites—day and night, light and shadow, life and death.

In Hindu mythology, Mercury is known as **Budha**, a god associated with intellect, wisdom, and eloquence. Budha, like Hermes and Mercury, governs communication and trade, emphasizing the planet's universal role as a symbol of movement, adaptability, and thought.

Astrologically, Mercury has long been regarded as the planet of communication, intellect, and travel. Its swift orbit around the Sun, completing a year in just 88 Earth days, mirrors its association with speed and dynamism. In horoscopes, Mercury's position is believed to influence how people think, speak, and connect with others. Whether or not one accepts astrology, the cultural significance of Mercury as a force of thought and interaction remains deeply ingrained.

Beyond its symbolic associations, Mercury also embodies the idea of transformation. In alchemy, Mercury was a central figure, representing the element quicksilver (liquid mercury), which could flow and adapt to any container. Alchemists viewed mercury as both a literal substance and a metaphor for transformation, a bridge between physical matter and spiritual enlightenment. The planet Mercury, elusive and ever-changing, fits this metaphor perfectly—a reminder of the impermanence and fluidity of existence.

Mercury's mythological associations are not merely historical curiosities; they are reflections of how humans have always sought to understand the natural world. By observing Mercury's movements, ancient cultures wove stories that connected the celestial with the earthly, creating a bridge between the cosmos and the human experience.

Mercury

Even today, Mercury's mythology persists in the way we think about the planet. We continue to associate it with speed, communication, and adaptability—qualities that resonate in an age defined by rapid change and constant movement. Its name, borrowed from a messenger of the gods, serves as a reminder of the connections that bind us, not just to each other but to the universe itself.

The mythology of Mercury also challenges us to see the planet as more than a physical object. It invites us to consider its symbolic significance, its role as a reminder of balance, duality, and transformation. Mercury is a world that exists at the extremes, yet it endures, moving swiftly and gracefully through its orbit.

Through its myths, Mercury becomes more than a planet; it becomes a metaphor for the human condition—a symbol of our own journeys, the connections we forge, and the transformations we undergo. It is a reminder that even the smallest and most fleeting things can hold profound meaning, that in motion and change, we find the essence of life itself.

Mercury

Chapter 7: Life at the Extremes

Mercury, with its searing days, frigid nights, and lack of atmosphere, seems an unlikely candidate for life. On the surface, it appears to be the antithesis of a habitable world—devoid of water, shelter, and the moderating forces that make Earth so uniquely suited to sustaining life. Yet, as our understanding of biology and planetary science evolves, so too does our sense of what is possible. Could Mercury, in its extremes, harbor life in some form?

To begin with, Mercury's surface is one of the harshest environments in the solar system. Daytime temperatures on the sunlit side can reach an astonishing 430 degrees Celsius, while the night plunges to a frigid minus 180 degrees. With no atmosphere to mediate these swings, the surface experiences the full force of the Sun's radiation and the unyielding cold of space. Under such conditions, life as we know it—dependent on liquid water, moderate temperatures, and a stable environment—would struggle to survive.

Yet there is more to Mercury than its hostile exterior. In the permanently shadowed craters near the planet's poles, conditions are vastly different. These regions, shielded from the Sun's rays by steep crater walls, remain in perpetual darkness, with temperatures plunging low enough to preserve water ice. While the ice is frozen solid and mixed with organic compounds, it represents a stable environment, free from the extreme heat that dominates the rest of the planet.

Could these polar ice deposits provide a haven for extremophiles, organisms capable of surviving in harsh environments? On Earth, extremophiles thrive in places once thought inhospitable: hydrothermal vents at the ocean floor, acidic hot springs, and even the frozen tundra of Antarctica. Some microorganisms, such as tardigrades, can survive extreme heat, freezing cold, radiation, and even the vacuum of space. These lifeforms demonstrate the resilience of biology and suggest that life's boundaries are far wider than we once imagined.

While Mercury's polar craters are among the most stable environments on the planet, they would still pose significant challenges to life. The water ice, while present, is mixed with carbon-rich compounds that could be toxic. Furthermore, the lack of atmosphere means there is no protection from cosmic radiation, which bombards the surface continuously. Any potential life in these regions would need to either shield itself beneath the ice or possess mechanisms to repair radiation-induced damage.

The potential for life on Mercury also raises questions about its origins. If extremophiles exist on Mercury, are they native to the planet, or could they have arrived from elsewhere? Panspermia, the idea that life can travel between planets on meteorites, provides a possible explanation. Impacts from comets and asteroids could have delivered organic compounds and microbial life to Mercury's surface, depositing them in the shadowed craters where conditions are most favorable for preservation.

Another intriguing possibility is that Mercury's early history may have been more conducive to life. Billions of years ago, when the solar system was young and chaotic, Mercury may have retained a thicker atmosphere and experienced volcanic activity that released water vapor. Under these conditions, liquid water could have existed on its surface or beneath it, creating temporary environments where life might have emerged. If so, the polar ice deposits could hold chemical traces of this ancient life, preserved for eons in the perpetual cold.

The search for life on Mercury is not just a scientific exercise; it is also a philosophical one. It challenges us to reconsider what it means for a planet to be "alive." Must life, as we search for it, conform to the conditions we know on Earth, or can it exist in forms and environments that are entirely alien? Mercury's extremes invite us to expand our definitions, to imagine life as an adaptable, creative force that can find a foothold even in the most unlikely places.

Beyond the question of life itself, Mercury offers lessons about survival in the face of adversity. Its surface, battered by impacts and stripped of atmosphere, is a testament to resilience. The polar craters, with their water ice and organic compounds, show how stability can exist even in the midst of chaos. In this way, Mercury mirrors the principles that define life: persistence, adaptation, and the ability to endure against overwhelming odds.

Mercury

Future missions, such as the **BepiColombo spacecraft**, will provide critical data about Mercury's polar regions, including detailed maps of the ice deposits and measurements of their composition. While the discovery of life on Mercury remains a distant possibility, these missions will deepen our understanding of the planet's potential to support extremophiles and the broader conditions that allow life to exist in extreme environments.

Even if Mercury ultimately proves lifeless, its extremes are not without meaning. They serve as a reminder of the vast range of possibilities within the universe and the resilience of matter in the face of adversity. In Mercury, we see the boundaries of survival pushed to their limits, a world that challenges our assumptions about habitability and the adaptability of life.

The idea of life on Mercury, improbable as it may seem, forces us to confront the limits of our imagination. It reminds us that the search for life is not just a quest for answers but a journey into the unknown, where every discovery expands our understanding of what it means to exist.

Mercury, in its starkness and extremes, is a testament to the universe's capacity to surprise us. Whether or not it harbors life, it remains a symbol of endurance, a world that has persisted in the face of the Sun's relentless power.

Mercury

Chapter 8: Lessons from a Fiery Neighbor

Mercury's existence, so close to the Sun's blinding glare, is a paradox of survival. It is a world that endures extremes of heat and cold, a planet of barren landscapes and improbable ice, and a witness to the raw power of the solar system's most dominant force. While Earth flourishes in a temperate orbit, Mercury exists on the edge of destruction, locked in a dance with a star that both sustains and threatens it. In its precarious position, Mercury offers lessons—not just scientific, but philosophical—about resilience, adaptation, and the nature of existence under overwhelming forces.

One of Mercury's most profound lessons lies in its ability to endure. From its dense metallic core to its cratered surface, Mercury is a planet that has withstood billions of years of bombardment, solar radiation, and gravitational tug-of-war with the Sun. It is a testament to the persistence of matter in the face of cosmic forces that seem determined to unmake it. The vast Caloris Basin, the polar ice deposits, and the scarps that crisscross its surface are reminders that even in the most hostile conditions, there is the potential for stability and endurance.

This endurance is not passive; it is the result of Mercury's unique properties. Its immense iron core, which likely accounts for its density and magnetic field, acts as a stabilizing force, anchoring the planet against the gravitational pull of the Sun. Its lack of atmosphere, while rendering the surface uninhabitable, also preserves its features, allowing the planet to serve as a geological time capsule. In this way, Mercury demonstrates that survival is not merely a matter of resisting destruction but of adapting to one's environment, no matter how extreme.

Another lesson Mercury imparts is the value of simplicity. Stripped of the complexities that characterize other planets—lush atmospheres, flowing rivers, shifting tectonic plates—Mercury reveals the essence of planetary existence. Its surface tells a story of impacts and cooling, contraction and preservation. In its starkness, Mercury becomes a mirror for the

Mercury

fundamental forces that shape all worlds, forces that are often obscured by the more dynamic processes at work on planets like Earth and Venus.

Mercury also teaches us about the power of proximity. Its closeness to the Sun makes it a laboratory for understanding the effects of intense solar radiation and gravitational forces. The tidal locking that slows Mercury's rotation, the asymmetry of its magnetic field, and the volatile loss from its surface are all direct consequences of its relationship with the Sun. These phenomena remind us that proximity to power—be it physical, cosmic, or metaphorical—comes with both opportunities and costs.

Philosophically, Mercury's existence challenges our assumptions about balance and survival. It operates at an extreme, orbiting closer to the Sun than seems reasonable, yet it persists. Its ability to retain polar ice, its capacity to generate a magnetic field, and its role as a witness to the early solar system are testaments to the universe's ability to create equilibrium in unlikely places.

This balance is not without sacrifice. Mercury's proximity to the Sun has cost it much: a thicker atmosphere, a more temperate environment, and the dynamic processes that characterize other terrestrial planets. Yet, what Mercury lacks, it compensates for in resilience. It has found a way to exist, not in spite of its proximity to the Sun but because of it.

Mercury's lessons extend beyond planetary science to questions about humanity's place in the universe. Like Mercury, we exist in a delicate balance, orbiting a star that sustains us but could also destroy us. We rely on the Sun's energy for life, yet we are vulnerable to its whims, from solar storms that disrupt communication systems to the long-term threat of climate shifts influenced by solar variability.

Mercury reminds us that survival is not about avoiding extremes but navigating them. It challenges us to consider how we adapt to our own proximity to overwhelming power, whether that power is natural, technological, or social. How do we, like Mercury, find stability in the face of forces far greater than ourselves? How do we preserve what matters while accepting the inevitability of change?

Mercury

In its starkness, Mercury also invites us to consider the beauty of extremes. Its surface, pocked with craters and etched with cliffs, is a canvas of cosmic history. Its polar ice, gleaming in perpetual shadow, is a paradox of preservation amid heat. These features, born of Mercury's proximity to the Sun, remind us that even the harshest environments can produce moments of wonder.

Mercury's role as a fiery neighbor also offers a broader perspective on resilience. It is not a world designed for life, yet it endures. Its very existence challenges the boundaries of what is possible, showing that survival is not a matter of ideal conditions but of adaptation and persistence.

As we look to the stars and explore planets beyond our own, Mercury serves as a guide, teaching us to appreciate the diversity of worlds and the resilience of matter. It reminds us that even in the shadow of immense power, there is room for complexity, stability, and beauty.

Mercury's lessons, like the planet itself, are stark but profound. It is a world that speaks to the extremes of existence, the balance between creation and destruction, and the resilience that defines life at all levels. In its orbit so close to the Sun, Mercury reflects the paradoxes of survival, teaching us that even in the most unforgiving environments, there is something to be learned, something to endure, and something to marvel at.

Conclusion: Mercury's Paradox

Mercury is a planet defined by contradictions. It is the closest to the Sun, yet its polar regions hold water ice, preserved in perpetual shadow. It is the smallest planet in the solar system, yet its density rivals that of Earth. It is battered, barren, and seemingly lifeless, yet its story is one of resilience and survival. To study Mercury is to grapple with paradoxes, to see how extremes coexist and how a world can persist at the very edge of possibility.

Mercury's very existence challenges our assumptions about what a planet should be. It has no weather to shape its surface, no flowing rivers or shifting tectonic plates to renew its crust. It lacks the thick atmospheres of Venus and Earth, the volcanic drama of Io, or the subsurface oceans of Europa. Yet, in its stark simplicity, Mercury tells a story as rich and complex as any other world in the solar system.

One of Mercury's most profound lessons is its endurance. Born in the chaos of the early solar system, it has survived billions of years of impacts, solar radiation, and gravitational pull. Its surface, pockmarked with craters and crossed by immense scarps, bears witness to the violence of its past. Its dense core, mysterious magnetic field, and polar ice deposits reveal a planet that continues to evolve, defying expectations at every turn.

Mercury's proximity to the Sun places it in a realm of extremes, where the ordinary rules of planetary science often do not apply. Its slow rotation creates day-night cycles that last the equivalent of 176 Earth days, subjecting the surface to prolonged periods of searing heat and freezing cold. Its thin exosphere offers no protection from the solar wind, leaving the planet exposed to the full force of the Sun's radiation. And yet, Mercury persists, adapting to its environment in ways that challenge our understanding of planetary dynamics.

At the same time, Mercury's starkness offers clarity. Stripped of the atmospheric complexities of Venus and Earth, Mercury provides a glimpse into the raw forces that shape planets. Its surface preserves a record of the solar system's early history, a time when collisions and chaos dominated. Its core and magnetic field offer clues about the processes that sustain planetary dynamos, while its polar ice raises questions about the distribution of water and volatiles across the inner solar system.

Philosophically, Mercury's paradoxes invite reflection. It is a planet of extremes, yet it is stable. It endures despite its exposure to overwhelming forces. It is not a world of life, yet it holds water ice, the raw material for life. In Mercury, we see a microcosm of the universe's capacity for balance, resilience, and transformation.

Mercury also serves as a reminder of the interconnectedness of the solar system. Its composition reflects the conditions of the protoplanetary disk, its magnetic field offers insights into planetary cores, and its ice deposits hint at the delivery of volatiles by comets and asteroids. To understand Mercury is to understand not just a single planet but the forces that have shaped the entire solar system—and by extension, our own planet and existence.

As humanity continues to explore the cosmos, Mercury remains a guide and a challenge. It teaches us to expect the unexpected, to look for meaning in simplicity, and to see extremes not as barriers but as opportunities for discovery. Its paradoxes are not contradictions to be resolved but truths to be embraced, reflections of the universe's boundless creativity.

Mercury is more than the closest planet to the Sun. It is a survivor, a witness, and a teacher. Its lessons are as enduring as the planet itself, offering insights into resilience, balance, and the possibilities of existence at the edge of extremes.

End Note: Mercury and Beyond

Mercury is a planet of lessons, extremes, and paradoxes—a world that thrives on the edge of the Sun's immense power. In its endurance, it mirrors the resilience of the universe itself. Mercury's story is not just one of survival in an inhospitable environment but of adaptation and balance, revealing the intricate dance between destruction and persistence that defines the cosmos.

Yet, Mercury's importance does not end with itself. Its lessons reverberate across the solar system and beyond, offering insights into the nature of other planets, moons, and exoplanets scattered throughout the galaxy. From the icy craters at its poles to its shrinking scarps and dynamic magnetic field, Mercury's features challenge our understanding of what is possible, forcing us to reconsider the boundaries of habitability, stability, and planetary evolution.

In studying Mercury, we also gain perspective on worlds that exist under similar conditions elsewhere in the universe. Many exoplanets discovered in recent years orbit perilously close to their stars, enduring radiation, tidal forces, and scorching heat. These so-called "hot rocky planets" share traits with Mercury, providing opportunities for comparison. Mercury becomes a laboratory, a reference point for understanding how planets form, survive, and transform in the most extreme conditions.

The exploration of Mercury is far from over. Missions like **BepiColombo**, now en route to the planet, will continue to expand our knowledge, probing its composition, magnetic field, and surface in greater detail. These missions are not just about Mercury—they are about understanding the processes that shape planets throughout the universe, including our own.

Mercury also reminds us of the fragility and resilience of life. While its surface appears devoid of life, its ice-filled craters and carbon-rich compounds hint at the shared history of the solar system, a history in which the ingredients of life are scattered across even the most unlikely worlds.

Mercury

As we turn our gaze outward, Mercury offers a moment of reflection. It challenges us to appreciate the diversity of planetary experiences and to see beauty in simplicity and endurance. It asks us to consider how life, matter, and meaning persist in the face of overwhelming forces, whether on a rocky planet near the Sun or in the farthest reaches of the galaxy.

Mercury is both a destination and a beginning. Its story is a chapter in the greater narrative of the solar system, a tale of creation, transformation, and survival. In studying Mercury, we not only unlock the secrets of a small, scorched world but also gain a deeper understanding of the universe and our place within it.

As the light of the Sun glints off Mercury's surface, it carries with it the lessons of a planet that endures against all odds—a fiery neighbor that reminds us of the resilience inherent in existence and the boundless possibilities of exploration.

www.ingramcontent.com/pod-product-compliance
Lightning Source LLC
Chambersburg PA
CBHW070944220526
45469CB00007B/2517